Benedito Rodrigues da Silva Neto

Medical Mycology - Volume 1

Benedito Rodrigues da Silva Neto

Medical Mycology - Volume 1

Superficial mycoses

ScienciaScripts

Imprint

Any brand names and product names mentioned in this book are subject to trademark, brand or patent protection and are trademarks or registered trademarks of their respective holders. The use of brand names, product names, common names, trade names, product descriptions etc. even without a particular marking in this work is in no way to be construed to mean that such names may be regarded as unrestricted in respect of trademark and brand protection legislation and could thus be used by anyone.

Cover image: www.ingimage.com

This book is a translation from the original published under ISBN 978-613-9-76436-5.

Publisher:
Sciencia Scripts
is a trademark of
Dodo Books Indian Ocean Ltd. and OmniScriptum S.R.L publishing group

120 High Road, East Finchley, London, N2 9ED, United Kingdom
Str. Armeneasca 28/1, office 1, Chisinau MD-2012, Republic of Moldova, Europe
Printed at: see last page
ISBN: 978-620-6-35730-8

Presentation

The series of books on Medical Mycology is the result of studies, classes, seminars and presentations prepared for undergraduate and postgraduate students and health professionals.

0 main objective is to consolidate knowledge and coordinate learning based on relevant scientific and literary productions on the subject. The way in which each mycosis was grouped and categorized was based on the general principle of fragmenting the content as much as possible so that it could be visualized and understood when added to other scientific publications.

According to scholars, there are believed to be around 100,000 species of fungi, of which approximately 150 are responsible for causing pathologies in humans and animals, ranging from systemic infections to cutaneous, subcutaneous or mucosal infections.

In fact, knowledge has expanded and the production of new data in the form of research results has been very constant in recent years, so the proposal of this material subdivided into seven volumes aims to add knowledge by uniting information already published and proposing to be a solid basis for understanding future results that will add to the knowledge of medical mycology.

Finally, como spokesperson for the new generation of medical mycologists, firstly, we aim to bring the spotlight back to the field of fungal diseases, attracting the attention of the public health spheres to the important and considerable pathologies that affect the population and continue to be neglected. Secondly, through a didactic and enjoyable read, it will attract new and promising academics to the field of mycology research.

Summary

INTRODUCTION

We consider mycology to be the study of microorganisms that have yeast-like and/or filamentous aspects, known as fungi. It is therefore a wide-ranging field of study that involves many researchers in different scientific, technological and industrial fields. In this series of books, we will focus exclusively on the aspects inherent in the interference of these fungi in human health.

From Whittaker's description in 1969, based on morphology and mode of nutrition, the Fungi kingdom was described and in 1990 grouped into the *Eukaria* domain by Carl Woese, who also defined two other domains: *Archaea* and *Eubacteria.*

From 1839, with the studies of Schoenlein, Langenbeck and Gruby, on superficial mycoses, we can say that there was an indication of the first studies of what we know today as the field of medical mycology. In 1842 there were initial studies on Mycetomas with Gill, followed by the first studies on Aspergillosis in 1856, arriving in the field of dermatology with Raymond Jacques Adrien Saboraud, one of the most important publications in the field: LES TEIGNES, 1910.

However, it's important to note that there was a gap of almost 50 years until Saboraud's publication, which was due to the emergence of discoveries in the field of bacteriology with the appearance of great names such as Louis Pasteur, Robert Koch and Emile Roux. Following Saboraud, in 1934 Chester W. Emmons also emerged with the proposal to standardize medical mycology.

Even at the beginning of medical mycology, we had researchers like Rudolf Virchow correlating human medical practice with deep mycosis (1856). In 1894, Alejandro Posadas made the first observations of *Coccidioides immitis,* which was only identified as a dimorphic fungus in 1905 by Ophuls and Moffit.

0 Brazil is a benchmark when it comes to studies in medical mycology, both for its pioneering researchers and for the new generation armed with biotechnological and molecular developments. The fathers of Brazilian medical mycology include Adolf Lutz in 1908, followed by Alfonso Splendore and Floriano Paulo de Almeida in the identification of *Paracoccidioides brasiliensis, as* well as Alberto Thomaz Londero, Olga Fischman Gompertz and, above all, Professor Carlos da Silva Lacaz with his "Tratado de micologia medica" in 2002.

Medical Mycology, defined as the science responsible for studying diseases caused by fungi in both humans and animals, is becoming increasingly important due to the emergence of antibiotics that leave patients more vulnerable to infections, new and more effective diagnostic methods, the emergence of AIDS and immunosuppressive therapies.

Considering that the balance between environment, pathogen and host is fundamental to the definition of health, the probable imbalance in this triad is the focus of study in medical mycology, from the individual's work or living environment where samples can be collected to the direct visualization of the fungus in material collected directly from the lesion or from this same material kept in culture. The laboratory, an important key in this process, can lead us to immunological, metabolic and currently molecular findings in samples. Since we need to associate clinical and laboratory data, we will pay specific attention to each of these throughout this book, considering the varied and presumptive clinical criteria and at the same time taking into account advances in laboratory biotechnology.

The classification of mycoses can be based on the site of the mycosis, the route of infection and the virulence of the pathogen (BARON, 1996).

According to their location, mycoses can be classified as superficial, cutaneous, subcutaneous and systemic (deep). As for classification based on the route of infection, mycoses can be endogenous or exogenous. Endogenous mycoses are those in which the etiologic agent was part of the patient's normal microbiota or there was a reactivation of a previous latent infection. Exogenous mycoses are those in which infection occurs through inhalation, traumatic inoculation or contact. Finally, according to the virulence of the pathogen, we classify pathogenic fungi as those capable of causing infection in immunocompetent individuals. Opportunistic fungi, on the other hand, require an immunosuppressed host to cause disease and generally cause systemic mycoses.

In this first volume we intend to deal with mycoses of medical interest that develop alterations only in the most superficial layer of the stratum corneum and do not, in most cases, induce any inflammatory response in the host. These are called coto superficial or strict superficial mycoses.

Fungi are eukaryotic organisms, unicellular Qeveduriformes) or multicellular (filamentous), haploid [homo- or heterokaryotic], with a cell wall containing chitin and alpha-glucan. They have no plastids or photosynthetic pigments.

The term mycelium is often used to exemplify a cluster or group of fungal cells. Multicellular mycelium can be defined as a set of hyphae or filaments (whether septate or aseptate), while unicellular mycelium would be a set of budding yeasts.

Some fungi that normally present septate mycelium in the saprophytic phase, in nature or in laboratory cultures, when they pass into the parasitic phase in the human or animal organism, transform into simple rounded elements, reproducing by twinning, are the blastomycotic mycoses. A classic example in Brazil would be Paracoccidioidomycosis.

0 mycelium can be divided into two parts: vegetative mycelium (structure responsible for sustaining and absorbing nutrients), reproductive mycelium (structure responsible for reproducing and propagating the species).

Firstly, the vegetative mycelium, which can be unicellular, filamentous or pseudo-filamentous, only produces asexual propagules, since this structure does not have the purpose of reproduction, but its asexual structures also contribute to the propagation of the fungus. Asexual propagules include: Arthroconidia (spores formed by the simple dismemberment of septate hyphae. Together with the latter, they are used to diagnose Dermatophytoses (impingens, uric acid, chilblains, onychomycoses) in a skin scraping. It is the only type of spore found in the genus *Geotrichum* sp. Blastoconidium (a spore that forms by budding, normally found in yeasts) and o Chlamydoconidium (rounded structures that can be found along the filament or in the apical parts).

Secondly, we have the reproductive mycelium, which in turn is subdivided into asexual and sexual propagules that we haven't seen in the vegetative mycelium until now.

Asexual propagules can be found externally on the reproductive structure, called conidia, or internally, inside the reproductive structure known as the sporangium, and are then called sporangiospores.

Sexed propagules are also found in the external and internal forms of the fungal structure, but they have different names and this leads us to their classification in the four known classes of the Eumycophyta division: a) Zygomycetes (Phycomycetes) b) Ascomycetes c) Basidiomycetes d) Deuteromycetes or Imperfect Fungi, whose sexual reproduction is not yet fully known. Thus, we have external sexual propagules in which we find Basidiospores in the reproductive structure, and internal sexual reproductive propagules, both Ascospores and Zygospores.

Fungi can be found in the most diverse environments, from dry soil, rotting logs or on fruit, milk, water and dust. When they have a preference for soil, we call them geophilic, zoophilic for those found on animals and anthropophilic for those isolated from humans.

Most grow best between 25° and 30º C. However, this is not a limiting factor for the development of fungi, mainly because when in the parasitic state they prefer temperatures close to 37° C. The increase in the medium in which yeasts develop can also occur due to exothermic fermentation reactions caused by various enzymes such as saccharases, maltases, proteases, peptidases and even phosphatases, asparaginases etc.

Not only the enzymes mentioned above, but several others found in the cell wall, membrane, or those that are secreted into the extracellular environment have been the subject of study in molecular biology, genetic engineering and biotechnology.

Various fungal products are used commercially, from ethanol to antibiotics, pigments, vitamins and biological pesticides. In addition to becoming valuable research objects, in particular, como eukaryote models, since they are easily manipulated in the laboratory, provide important information on the biochemistry, genetics and molecular biology of eukaryotes.

An important point to be highlighted in medical mycology is the treatment of mycoses of medical importance. We currently have a limited number of possibilities for treatment, which is a negative point, but it does encourage new studies and lines of research that try to find new innovative treatment possibilities with fewer side effects for the patient.

Amphotericin B and the azole drugs (ketoconazole, fluconazole and itraconazole) have been the drugs of choice in therapy. The focus of these drugs is the membrane and the ergosterol pathway, increasing membrane permeability which allows the leakage of various small molecules, leading to cell death. Azole drugs cause fewer adverse reactions than amphotericin B, but are less potent than it. They can have fungistatic or fungicidal action. Unfortunately today, due to the excessive use of azois, we have widespread resistance in susceptible species.

Another systemic antifungal agent used is flucytosine. All sensitive fungi are able to deaminate flucytosine into 5-fluorouracil, a potent antimetabolite; as a result, their DNA synthesis is impaired.

Other antifungals also used are the morpholine derivative amorolfine; the thiocarbamates tolnaftate and tolcyclate; the allylamines naftifine, terbinafine and butenafine and the compound ciclopirox, as well as the antibiotic griseofulvin, which we will discuss and correlate at the appropriate time.

We can therefore briefly conclude that medical mycology has the main objective of establishing a mycological diagnosis of fungal infections, based on the correct collection and processing of clinical specimens. We will try to export each mycosis from the basic characteristics of etiology and habitat, through pathogenesis and clinic to observance of the rules of preservation, and proper transportation of clinical materials to processing sites, como Microbiology/Mycology laboratory demonstrating como a whole the importance and aspects of each mycosis.

GENERAL CONSIDERATIONS

Superficial mycoses are among the most frequent dermatological problems in Latin America, but because they are not included among the diseases that are compulsorily notifiable, their evaluation is complex and many of the data collected end up being somewhat incomplete.

In Brazil, authors have described superficial mycoses as a relevant part of skin diseases, since their prevalence is directly related to the population's economic level and favorable climatic conditions.

Superficial mycoses can be divided into three basic groups: Keratophytoses, Dermatophytoses and Yeast infections. Keratophytoses are exclusively superficial mycoses, which are located in the epidermal keratin and hair. The fungi that cause superficial mycoses are limited to the keratinized and semi-keratinized portions of the skin or its surface, hair and nails.

In this volume we will discuss keratophytoses, a specific group of superficial mycoses caused by various microorganisms that cause different types of dermatological lesions. So what is the common characteristic that leads us to group these mycoses together? The answer is: there are no known skin hypersensitivity reactions in this group, which occur in dermatophytoses and yeast infections.

We will therefore discuss the following mycoses that fall

under this heading: 1) Pityriasis versicolor 2) Ceratomycosis nigra or *Tinea nigra* 3) *Piedras* - white and black 4) Trichomycosis palmelina or Leptotrix 5) Erythrasma.

It's important to note here that o erythrasma and trichomycosis palmelina, caused by bacteria, are not mycoses, but have traditionally been studied in mycology, and precisely for this reason we have included them here como in the group of keratophytoses.

1. Miselanea: Pitiriasis Versicolor

I) DEFINING

A keratophytosis characterized by patches of skin with shades varying from hypo- to hyperpigmentation is called pityriasis versicolor.

Pityriasis versicolor is a chronic and benign dermatosis, which until now was considered to be the most common skin pigmentation disorder in the world. It manifests itself through hypochromic or brownish macules or, even less frequently, erythematous, oval and well delimited and covered with fine scales, these thin-looking scales are called furfuraceas. They begin as small perifollicular macules that tend to converge and form extensive, irregular lesions. It usually affects the trunk, upper limbs and face, but can sometimes be seen on the crural region, lower limbs, scalp and palmar and plantar regions.

In 1889 Baillon gave the fungus the name *Malasseziafurfur*, but since 1853 pityriasis versicolor has been considered a fungal infection. At first, the term *Malassezia* was used to characterize the mycelial phase of the fungus and *Pytirosporum* was used for the yeast phase. Nowadays we only use the term *Malassezia to* refer to both forms, and we now know that not only *Malassezia furfur* is the causative agent of *pityriasis* versicolor, but other species such as *M. pachydermatis, M. sympodialis, M. globosa, M. obtusa, M. restricta* and M. *slooffiae.*

Although the diagnosis is simple, as we'll see below, the

most complicated aspect of treatment is the likelihood of recurrence.

II) ETIOLOGY

Malassezia spp. is an etiologic agent comprising seven species: *M. furfur, M. pachydermatis, M. sympodialis, M. globosa, M. obtusa, M. restricta* and *M. slooffiae.* Over the last 100 years, other genus and species names have been created. However, the term "yeast of the genus *Malassezia"* has been established only for lipophilic fungi that are part of the normal microbiota of the skin.

III) HABITAT

Malasseziacomo, the agent of mycosis, is a normal inhabitant of human skin (anthropophilic), appearing much more frequently in seborrheic conditions.

IV) EPIDEMIOLOGY

It is prevalent in tropical and sub-tropical climates, but its scope is universal. Pityriasis versicolor is a universally distributed disease, more prevalent in tropical and subtropical climates, reaching a frequency of around 50% of the population in certain places and ethnic groups (GUPTA et al, 2002).

V) PATHOGENY

The pathogenesis is not yet fully understood. It seems certain, however, that it is of endogenous origin, with man normally living with its causative agent in a saprophytic state. It is caused by the overgrowth of lipophilic yeasts of the *Malassezia* genus, which are part of the normal microbiota of the seborrheic areas (scalp, face, medial portion of the trunk) of most adults. Environmental changes or immunological alterations in the host can facilitate an overgrowth of these yeasts with the development of lesions. Factors that increase the chance of pityriasis versicolor occurring are: increased activity of the sebaceous glands, which occurs in adolescence, the use of oily oils and creams, sun exposure and, above all, the seborrhea is the substrate that best adapts to the pityriasis versicolor parasite.

These yeasts are dimorphic and when they are part of the normal microbiota they are found in the form of yeasts. When parasitized, they are yeasts with short, septate and unbranched hyphae.

Malassezia metabolizes fatty acids by means of lipases and, among the metabolites released in this pathway, is o azelaic acid, which inhibits the action of the enzyme dopa-tyrosinase, o which blocks the conversion of tyrosine into melanin and сото consequently results in the appearance of hypochromic spots. There is no inflammatory infiltrate in the skin affected by these lesions (MENDEZ-TOVAR, 2010).

We know from the literature that antigenic components of fungi participate in the interaction with the host's immune cells. However, further studies are needed to elucidate the immunological and pathogenic processes that correlate

Malassezia with seborrheic diseases.

VI] CLINIC

To put it bluntly, what most often brings patients to the clinic are reasons related to aesthetic appearance, given the unpleasant aspect caused by the blemishes. Skin patches vary greatly in size and can cover a large area of the skin.

As already mentioned here, the characteristic location is on the trunk and upper limbs, face and neck, but it can also be seen on the palms of the hands or soles of the feet. Usually, pityriasis versicolor is a simple scaly patch, but it can take on a variety of dermatological aspects: erythematous-papulosis, papulosis, follicles, etc.

papulose, circled shape.

The shade of Colorado varies and can be hypo- or hyperpigmented, which is why we use the name versicolor. This variation can be explained both by the action of the sun, which causes it to warm up, and by the direct action of the parasite on melanogenesis. Several studies have tried to explain the variation in tone of the lesions, sometimes in the same patient. Hyperpigmented lesions seem to occur both due to an increase in the size of melanosomes and changes in their distribution in the epidermis, as well as an increase in the *turnover of* cells in the corneal layer. Hypopigmented lesions may be the result of the inhibition of the dopa-tyrosinase reaction by lipidic substances (dicarboxylic acids) produced by the fungus when it is in a fatty environment, causing little melanization (Zaitz, 1998). *In vitro,* the mechanisms responsible for hypopigmentation occur through the production of dicarboxylic acids, with an inhibitory effect on tyrosinase, and a cytotoxic effect on melanocytes (Nogueira et al, 1988).

Yeasts of the *Malassezia* genus can colonize the nails of individuals who have predisposing factors such as diabetes, nail psoriasis, seborrheic dermatitis, contact dermatitis, nail trauma or who are using immunosuppressive drugs, causing the so-called onychomycosis, with a yellowish-white nail plate and subungual hyperkeratosis with onycholysis, and no paronychia.

Finally, it is important to note that *Malassezia* can sometimes become systemic (deep-seated), producing malassezia in patients undergoing intravenous lipid feeding.

VII) DIAGNOSIS

0 diagnosis is predominantly clinical. Examination with Wood's lamp can be useful, as the lesions turn a greenish-yellow color. 0 direct examination is very classic and widely used: the sample is placed between the coverslip and laminate, with one or two drops of 10% lactophenol or potassium. 0 parasite is presented in the form of a rounded element, or more or less quadrangular, characteristically agglomerated in varying numbers from 5 to more than 30 elements. Alongside these grouped forms, we can see a variable number of short hyphae, which can reach more than 10 pm in length. Culture is not done

for diagnostic purposes, as the agent's culture is quite complex. However, if necessary, the skin scales should be sown on Saboraud agar medium with antibiotics and olive oil and incubated at 25°C. The growth of creamy yellow-white colonies is observed after about eight days.

The great importance of direct microscopic examination lies in the speed of diagnosis. Currently, Kaneko et al, in 2006, developed an identification *kit,* modified agar medium for *Malassezia* (CHROMagar) for nine *Malassezia* species *(M. furfur, M. slooffiae, M. sympodialis, M. restricta, M. obtusa, M. globosa, M. pachydermatis, M. dermatis* and *M. japonica)* based on their biological differences and confirmed by molecular biology.

VIII) TREATMENT

There are several alternatives, among the most commonly used:

For topical treatment, use keratolytic agents in the form of lotions, shampoos or soaps. Sodium hyposulphite, selenium sulphide and salicylic acid are the most commonly used. Imidazole derivatives (ketoconazole, clotrimazole, econazole, oxiconazole) cyclopirox olamine, morpholamine derivatives and terbinafine 1% (Leeming et al, 1997) are effective in topical therapy.

Systemic treatment is indicated with azole derivatives such as ketoconazole 200mg/day for 10 or more days or triazole derivatives: itraconazole 200mg/day for five days (Zaitz,

Sampaio, 1995) or fluconazole 400mg, single dose (Faergemann, 1992).

SUGGESTED BIBLIOGRAPHY

AZULAY, R. D.; MONTEIRO, E. & AZULAY, E. - Superficial mycoses; their frequency in Rio de Janeiro. An. Brasil. Derm., 42:91-6,1967.

ASSAF RR; WEIL ML - The Superficial Mycoses Dermatol. Clin. 14 (1):57-67;1996.

BARON S, editor, WHALSH TJ, DIXON DM. Medical Microbiology.4Th Edition. Texas, 1996.

CUCE LC; BELDA JUNIOR W; RIBEIRO EB. - Itraconazole in the treatment of pityriasis versicolor: comparison between 5 and 7 days of treatment. Rev. Inst. Med. Trop. Sao Paulo32 (3) : 181-4; 1990.

FAERGEMANN J. Treatment of pityriasis versicolor with a single dose of fluconazole. Acta Derm Venerol 1992; 72:74-5.

GUEHO E; MIDGLEY G; GUILLOT J. - The Genus Mallassezia with description of four new species Antonie van Leeuwenhoek (69): 337-355; 1996.

GUPTA AK, BLUHM R, SUMMERBELL R. Pityriasis versicolo. J Eur Acad Dermatol Venereol 2002; 16:19-33.

KANEKO T, MAKIMURA K, SUGITA T, YAMAGUCHI,H. Tween 40based precipitate production observed on modified chromogenic agar and development of biological identification kit for *Malassezia* species. Med Mycol 2006; 44:227-231.

GUEHO E; MIDGLEY G; GUILLOT J. - The Ge- nus Mallassezia with description of four new species An- tonie van Leeuwenhoek (69) : 337-355; 1996.

LACAZ, C. S. - Compendium of Medical Mycology. Sarvier Edit. Univ. Sao Paulo; 1967.

LEEMING JP, SANSOM JE , BURTON JL. Susceptibility of *Malassezia furfur* subgroups to terbinafine. Br J Dermatol 1997; 137:764-7.

MENDEZ-TOVAR LJ. Pathogenesis of dermatophytosis and tinea versicolor. Clinics in Dermatology. 2010; 28: 185-189.

NOGUEIRA MCM, COSTA RO, SUDO L, PORTO JA. *Malassezia furfur* (in Pityriasis versicolor and in healthy individualsj. An Bras Dermatol 1988;63(suppl.lJ:229-34.

ROBERTS, S.O.B. - Pityriasis versicolor, a Clini- cal and Mycological Investigation. Brit. J. Derm., 81 : 315; 1969.

ROBERTS, S.O.R. - Pityrosporum orbiculare. Incidence and Distribution in Clinically Abnormal Skin. Brit. J. Derm., 81 : 264; 1969.

ROBERTS, S.O.B. - Pityriasis versicolor, a Clinical and Mycological Investigation. Brit. J. Derm., 81 : 315; 1969.

SCHMUTZ JL; BARBAUD A; CONTETAU-DONNEAU N. - Superficial mucocutaneous mycosis. Rev. Prat. 46, 13 : 1617-22; 1996.

SHRUM JP; MILLIKAN LE; BATAINCH 0. - Superficial fungal infections in the tropics. Dermatol. Clin. 12,4: 687-93; 1994.

ZAITZ C. Superficial mycoses proper. In: Zaitz C, Campbell I, Marques AS , Ruiz LR, Souza VM . Compendium of medical mycology. Sao Paulo: Medsi, 1998. p.65-79.

2. Ceratomycosis nigra: *Tinea Nigra*

I] DEFINING

Another keratophytosis we can highlight is *Tinea nigra.* A keratophytosis that manifests itself by the appearance of spots with a black hue, in keeping with its name, sometimes not so dark in color (light brown), with a more characteristic distribution in the palmar region, although it can be found in any region of the cutaneous integument. Also called black keratophytosis, Keratomycosis nigricans palmaris, black pityriasis, Microsporosis nigra, Keratophytia nigra, epidermal cladosporiosis and Keratomycosis nigricans.

We characterize it as a demaceous fungus that can be isolated from moist environments such as soil, plants and beach sand.

II] ETIOLOGY

In 1921, Parreiras Horta described the fungus *Cladosporium werneckii.* Today, the accepted name is *Hortaea werneckii (Phaeoannellomyces werneckii; Exophiala werneckii). Stnella araguata has* also been isolated from cases of *tinea nigra.*

III] HABITAT

Several fungi with similar characteristics to *Hortaea werneckii* can be found widely distributed in nature. The fungus *Phaeoannellomyces werneckii,* is a saprobic fungus that inhabits the environment, mainly places with a high salt concentration como beach sands, soil, dirty water, decaying vegetables, humid environments and normal skin (scalp and interdigital spaces).

IV] EPIDEMIOLOGY

It is considered to be restricted to tropical and subtropical areas. Mycosis is chronic and can affect children, immunocompetent people and especially women. The vast majority of cases occur in Brazil and the East.

V] PATHOGENY

The mode of infection is still very obscure, although there are cases correlated with members of the same family. It can affect individuals of any age and from all social groups, however, it predominates in young people under the age of twenty, of white race.

VI] CLINIC

Asymptomatic black spots. The margins may be slightly raised and are slightly scaly. There is no itching, except for psychogenic reasons. The location is most often on the palms of the hands. However, cases have been observed on various other areas of the body. Although it is a superficial phaeohyphomycosis, the term *tinea nigra* is still used.

VII) DIAGNOSIS

Direct examination of the scraping of the lesions, between lamina and laminula, with lactophenol or 20% soda, can show more or less dark, septate, sinuous hyphae. You can also see rounded, twinning elements, up to 5 pm in diameter.

The culture can be done in Sabouraud's medium, where black, shiny colonies are observed, as if they were yeasts, and green-olivaceous hyphae appear after a few days. The uni- or bicellular conidia are born directly from the hyphae by twinning, or by means of a spicule.

It is worth emphasizing the importance of differential diagnosis, since *tinea nigra is* similar to melanoma. Melanoma lesions are indurated or raised with a red color present in the dark lesion, and o melanoma is unlikely to occur on the palm of the hand.

VIII) TREATMENT

Oral imidazoles can be used, but retinoic acid, tincture of iodine 1-2%, sulphur solution 3% and thiabendazole 10% in the form of a cream, for example, can also be used with the same results.

Whitfield's ointment with the composition Salicylic acid (1.8 g) Benzoic acid (3.6 g) Vaseline (60.0 g) has been used successfully in the vast majority of cases.

SUGGESTED BIBLIOGRAPHY

BELFORD, LACAZ E SAMPAIO - Tinea nigra palmaris. First cases in Sao Paulo. Rev. Paul. Med., 57 : 386-397; I960.

BONIFAZ A, GOMEZ-DAZA F, PAREDES V, PONCE RM. Tinea versicolor, tinea nigra, white piedra, and black piedra. Clinics in Dermatology. 2010; 28: 140-145.

COOKE, W.B. - A Taxonimic Study in the Black Yeasts. Mycop, et Myc.Appl.,17,1 :42;1962.

DE ALMEIDA JUNIOR HL; SALEBIAN A; RIVI- TH EA - Ultrastructure of black piedra. Mycoses, 34, 11-12 : 447-51; 1991.

FIGUERAS MJ; GUARRO J; ZAROR J. - Ultrastructural aspects of hair digestion in black piedra infec- tion. J. Med. Vet. Mycol. 35, 1 : 1-6; 1997.

LACAZ, C. S. - Compendium of Medical Mycology. Sarvier Edit. Univ. Sao Paulo; 1967.

PARREIRAS HORTA - On a new case of Tinea preta and a new mushroom *(Cladosporium werne- ckii* Horta, 1921). Rev. Med. Cir. Br" 29 : 269-274; 1921.

REVANKAR, S G; SUTTON, D A. Melanized fungi in human disease. Clin Microb Rev 2010; 23 (4):884-928.

STONE,O.J.ETAL.-ThiabendazoleinDi- methylsulfoxide for Tinea nigra palmaris. Arch. Derm., 93, 2 : 241-242; 1966.

3. Nodular trichomycoses: *Piedras* - white and black

I) DEFINING

Stones or *Piedras* are all mycoses that form hard, stone-like nodules on hair and fur. Their color can vary from dark or black to white or yellowish-white.

II) ETIOLOGY

Trichosporon, a group of yeasts represented mainly by the species *T. cutaneum, T. Inkin, T. Ovoides* and *T. beigelii,* is responsible for *white piedra,* while *Piedraia hortae* develops the black variety.

III) HABITAT

Most *Trichosporon* species have been isolated from humans and a few from wood. *Piedraia hortae* has also not yet been isolated in the wild, although everything indicates that it lives in warm, humid areas of equatorial and tropical zones.

IV) EPIDEMIOLOGY

The characteristics of *piedras* are very similar in terms of epidemiology. They are prevalent on the African and American continents and are practically non-contagious. White *piedra* is infrequent, black *piedra* falls under what is known as superficial phaeohyphomycosis.

V) PATHOGENY

White *stone* is generally asymptomatic and not very contagious. The source of infection has been described as the microbiota of the skin or intestine, but as we have already said, these fungi are found in various microenvironments in nature. According to some researchers, white piedra can cause hair loss, this being related to the apparent ability of yeasts of the *Trichosporon* genus to degrade keratin, causing it to break down, but other authors suggest that this infection only affects the hair cuticle.

Black *piedra* caused by the dematiaceous *Piedraia hortae* does not affect the skin or the area around the hair. The main source of contamination, according to the literature, is soil, and it is also known that black *piedra* is transmissible, at least among children. In the case of genito-crural white stone, a certain lack of hygiene, sweating and friction are also factors linked to pathogenesis.

VI] CLINIC

White *pityriasis* has the clinical appearance of nodosities in the middle of the hair, which are yellowish-white or brownish and fusiform. It is worth remembering that *Trichosporon* species can also be related to onychomycosis, and in reports of deep infections affecting immunocompromised patients.

Unlike white *piedra*, black piedra is strongly adhered to the hair and is not as easy to remove as white *piedra.*

VII] DIAGNOSTIC©

0 direct examination between lamina and laminula, in lactophenol or 20% potash, immediately reveals a nodule. The black stone, in addition to its coloring, shows the presence of clear spaces, which are ascigerous stores (oval-shaped nodules). In the white stone nodule, in addition to the clear coloration, the ascigerous stores are missing. Another nodular trichomycosis that could be confused with stones is axillary trichomycosis, but we'll talk about it later and note that its bacterioid elements are quite distinct.

In terms of culture, *Trichosporon* sp. has colonies with a cereal-like, cerebriform appearance, covered with a layer of whitish hyphae and *Piedraia hortae* grows сото a dark, black/green colony, which can be raised or flattened in the central part.

VIII) TREATMENT

Clotrimazole can be used, but in general the affected hairs are shaved off and a solution of sublimate at 1/1000 or even 1/2000 in 60% alcohol is used. Itraconazole has recently been shown to also have great potential for treating *stones.*

SUGGESTED BIBLIOGRAPHY

AVRAM A; BUOT G; BINET 0; GRACIA AM; CESARINIJP - Clinical and mycological study of 11 cases of genitopubic trichosporosis nodosa (white pie- dra). Ann. Dermatol. Venereol. 114, 6-7 :

81927; 1987.

CARNEIRO, J.A. E ARAUJO, F.A. - New cases of Genital White Piedra. IV Brazilian Congress of Microbiology - Niteroi; 1972.

DU BOIS - Trichosporon isolated from pubic hair. Ann. Derm. 447; 1910, quoted in Simons.

FONSECA, O. E AREA LEAO - About Brazilian stone mushrooms. Supplement to Mem. Inst. 0. Cruz, 4 - December; 1928.

LONDERO AND FISCHMAN - On two cases of white piedra on the scalp in Rio Grande do Sul. Rev. Inst. Med. Trop. Sao Paulo, l,4:260;1959.

LACAZ, C. S. - Compendium of Medical Mycology. Sarvier Edit. Univ. Sao Paulo; 1967.

NINO, F. - Contribuicion al estudo de las Trico- patias piedricas en Venezuela. Mycop, et Myc. Appl. 2, 5-12; 1939.

VIEDRAS, A. P. - Notes on three Brazilian fungi. Piedrai hortai. Bragantia (Campinas), 3, 3:37 - 44; 1943.

ZELEDON, R. E V.M. DIAS - Considerations on a case of Piedrai tricosporica perineo scrotalis, without epidermal lesions. Hosp., 44,4:751-762;1953.

4. Trichomycosis Palmellina or Leptotrix

I] DEFINING

Nodular, Axillary or Chromatic Trichomycosis is a keratophytosis most often located in the axilla and sometimes in the pubis. Depending on the processes of association with other microorganisms, the nodules can become red or black in color.

In 1911 Castellan first described the parasite as *Nocardia tenuis.* Only in 1952 did Crissey and colleagues conclude that the causative agent of this trichomycosis was actually a diphtheroid *Corynebacterium tenuis.*

II] ETIOLOGY

Corynebacterium tenuis can be associated with *Micrococcus castellani* or *Micrococcus nigrescens.*

III] HABITAT

It is mainly located in the axillary region, with human skin being an ideal environment for its development.

IV) EPIDEMIOLOGY

Tropical, subtropical and temperate climates.

V] PATHOGENY

O *C. tenuis* does not seem to cause any damage to the hair, as washing with soap and water can remove the yellowish-white color. It is known that excessive sweating, lack of grooming, hot and humid climates facilitate infection.

VI] CLINIC

The nodules can vary in size and can join together to form a continuous mass around the hair. The colored varieties are o which in some cases can cause doubts.

VII] DIAGNOSTIC©

0 direct examination is the best mechanism for differentiating it from *piedra*. 0 palmelina trichomycosis nodules are made up of very short bacillary elements, which are easily seen when crushed between slides with lactophenol. Gram staining is recommended, as they are Gram positive.

On direct examination, we can also see color differentiation (flava variety, rubra variety and nigra variety). All the variations occur when *Corynebacterium tenuis* is associated with other pigment-producing bacteria of the genus *Micrococcus-. M. castellanii* and *M. nigricans.*

VIII) TREATMENT

Considered simple, the treatment mainly involves shaving the hairs and applying 1/1000 mercuric chloride in 70% alcohol, 2% formaldehyde in 70% alcohol, 1% lye or washing with soap and water because, as we said, the removal process is easier.

SUGGESTED BIBLIOGRAPHY

ASSAF RR; WEIL ML - The Superficial Mycoses Dermatol. Clin. 14 (1) : 57-67; 1996.

CASTELLANI, A. - Trichomycosis flava, nigra and rubra. Brith. J. Derm.,23:341;1911.

CRISSEY ET AL. - Study on the causative agent organism of Trichomycosis axillaris. J. Invest. Derm., 19 : 187-197; 1952.

LONDERO AND FISCHMAN - On two cases of white piedra on the scalp in Rio Grande do Sul. Rev. Inst. Med. Trop. Sao Paulo, l,4:260;1959.

HILDICK-SMITH ET AL. - Fungus Disease and Their Treatment. Little Brown and Company - Boston; 1964.

LACAZ, C. S. - Compendium of Medical Mycology. Sarvier Edit. Univ. Sao Paulo; 1967.

McCARTHY, L. - Tropical Mycosis. J.A.M.A., 123 : 459; 1943.

MONTEZ ET AL. - Electron microscopy study of infected hairs in Trichomycosis axillaris. J. Invest. Derm., 40,6: 277-278; 1963.

SCHMUTZ JL; BARBAUD A; CONTETAU- DONNEAU N. - Superficial mucocutaneous mycosis. Rev. Prat. 46, 13 : 1617-22; 1996.

SIMONS, R.G.D. - Medical Mycology. Elsevier Publishing Company - Amsterdam; 1954.

5. Erythrasma

I) DEFINING

The ultimate keratophytosis, which manifests itself as finely scaling erythematous plaques, preferentially located in the genito-crural region, then in the axillary region, interdigital spaces *and,* a few times, in various regions of the body, is known as Erythrasma.

II) ETIOLOGY

As we have already pointed out, the keratophytosis mentioned here is no longer considered fungal, but bacterial, with the etiologic agent being the diphtheroid *Corynebacterium minutissimum,* replacing *Nocardia minutissima.*

III) HABITAT

Anthrophilic, but with some findings in large mammals.

IV) EPIDEMIOLOGY

Predominantly found in hot and humid climates, but universal.

V) PATHOGENY

An interesting aspect of this infection is that the lesions fluoresce red under Wood's light.

It is often associated with diabetes. The disease manifests primarily in the folds of the body (groin, between the testicles, under the breasts, between the toes and in the armpits), where there is excessive sweating or poor hygiene. In some patients, the infection spreads to the trunk and anal region.

VI) CLINIC

In addition to the scrotal folds and thighs, other sites affected by erythrasma can be the axillary and pubic regions and the breast folds. The erythematous lesions are dry, irregular in shape and finely scaly. The lesions are reddish-orange at first, later turning a darker shade of brown. It can cause lamellar desquamation when seen in the generalized cutaneous form. Recently, findings in interdigital regions have received more attention.

VII) DIAGNOSIS

Starting from the area of the inguinal-crural region, which is usually also infected with dermatophytosis, yeast infections and pityriasis versicolor, we eliminated all three hypotheses by examining them fresh, between coverslip and laminate, with 20% potassium or lactophenol. We then proceeded with the GRAM test in order to observe numerous rod-shaped forms (coccoids and bacillus) stained in violet; we also observed granular bacterial filaments of varying lengths. Culturing the parasite in artificial media requires the presence of fetal bovine serum in the culture medium at 37° C.

VIII) TREATMENT

The best treatment requires Erythromycin, 250 mg, every 6 hours, despite the various commercial formulations already on the market.

SUGGESTED BIBLIOGRAPHY

ASSAF RR; WEIL ML - The Superficial Mycoses Dermatol. Clin. 14(1):57-67;1996.

GONCALVES, A.P. E G. MANGEON - Multiple erythrasma and non-intertriginous extension. Trab. Soc. Port. Derm. Sif., 18 : 11; I960, quoted by Hildick-Smith.

HILDICK-SMITH ET AL. - Fungus Disease and Their Treatment. Little Brown and Company - Boston; 1964.

LACAZ, C. S. - Compendium of Medical Mycology. Sarvier Edit. Univ. Sao Paulo; 1967.

McCarthy, L. - Tropical Mycosis. J.A.M.A., 123 : 459; 1943.

MUNRO-ASHMAN, WELLS AND CLAYTON (Erythrasma and adolescence). Brit. J. Derm., 75,10: 401- 404; 1963.

RIEGEL P; RUIMY R; RENAUD FN; FRENEY J; PREVOST G; JEHL F; CHRISTEN R; MONTEIL H. - Corynebacterium singulare sp. nov., a new species for urease-positive strains related to *Corynebacterium minu- tissimum.*

SIMONS, R.G.D. - Medical Mycology. Elsevier Publishing Company - Amsterdam; 1954.

SOMERVILLE, D. et al. - Erythrasma in a Hospital for the Mentally Subnormal. Brit. J. Derm., 82 : 355- 360 (1970).

From the material produced in this first volume, we aim to characterize and break down superficial mycoses and their main agents as clearly as possible.

Even though we are dealing with mycoses that at first will not develop skin hypersensitivity, this group of kerafitoses has become more relevant in epidemiological studies because they are correlated with other infections and also because of the search for well-being and body aesthetics.

In view of the importance of these mycoses known as keratofitoses, we will now explain their main characteristics and what differentiates them from other superficial mycoses.

This volume is part of a series of seven books, each with a specific focus on a particular mycosis of medical importance.

Printed by Books on Demand GmbH, Norderstedt / Germany